THE LOBSTER:
A MODEL FOR TEACHING
NEUROPHYSIOLOGICAL CONCEPTS

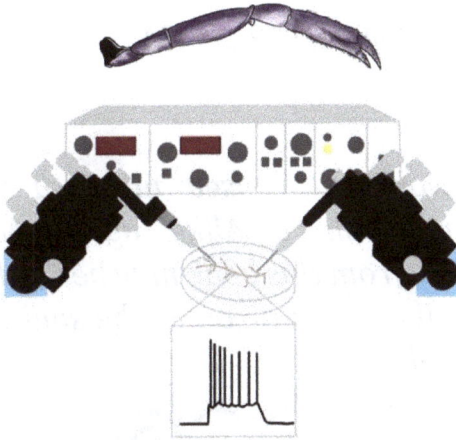

Kaitlyn E. Brock
and Robin L. Cooper

Azalea Art Press
Sonoma . California

ISBN: 978-1-943471-84-3

Cover illustration by:
Kaitlyn Brock and Alaina C. Taul

CONTENTS

ABSTRACT

The frog sciatic nerve remains a common educational model for animal and neurophysiology laboratories. It is used to teach principles of nerve conduction and related biophysical properties. However, animal care regulations and maintenance required for vertebrates pose some challenges. An alternative preparation used to teach the same principles as with the sciatic nerve may be beneficial.

The lobster preparation allows for the extraction of 10 to 20 nerves from a single animal. Lobsters are invertebrates, meaning there are fewer regulatory restrictions. They are relatively easy to house. This model would reduce the number of animals euthanized as a laboratory-based model for neurophysiology. Similar to the sciatic nerve, the lobster nerves can be used for teaching about conduction velocity, examining pharmacological actions, addressing recruitment of neurons, obtaining compound action potentials, and covering the principles of refractory periods. Lobster nerve preparations are relatively hardy and viable for hours at room temperature. They can be maintained for days if stored in a refrigerator.

This report explains how to obtain lobster nerve preparations and record from them for educational purposes. In addition, it demonstrates the recording of synaptic transmission at neuromuscular junctions and measures sensory proprioception all within a single animal.

TERMS

Compound action potential (CAP) - an extracellular recording of a summated electrical event from a nerve in response of a population of neurons.

Spike - Extracellular recording of an action potential from a nerve.

EJP - excitatory junction potential recorded in a muscle.

IJP - inhibitory junction potential recorded in a muscle.

PREFACE

This text is presented to provide this protocol for teaching neurophysiological concepts using the lobster as a model to anyone in the world. This resource will be open and free under a Creative Commons Share Alike (CC BY-SA) license, or at a low cost to download. This is a test run on inexpensive publishing of teaching protocols for our research group. If it proves to be of interest, additional educational protocols will be forthcoming.

This book is intended for instructors to use and modify as needed. Not every detail is spelled out. It is assumed that instructors know how to perform basic techniques of a physiology laboratory such as using extracellular and intracellular amplifiers, Faraday cages, and vibration-free tables as well as making sharp intracellular electrodes. There are citations for various, free online protocols published by video journal JoVE:

https://www.jove.com/journal.

They are very detailed in regard to some aspects of the protocols mentioned in this text.

We hope other researchers and educators will follow suit in publishing for low cost to others. In the meantime, there are a number of protocols and educational content available on the University of Kentucky website:

https://bio.as.uky.edu/users/rlcoop1.

Parts of the recordings highlighted were performed in a laboratory setting to obtain student comments and suggestions. There is a video provided on YouTube of steps which complement the text at:

https://www.youtube.com/watch?v=RvVMsekpgfU.

This protocol will also be presented at a neuroscience educational forum (Neuroscience Teaching Conference, July 17-19, 2024 at Wake Forest University, Winston-Salem, NC., USA) in order to obtain feedback on the protocol and content.

INTRODUCTION

Knowledge of electrical excitation and conduction in single neurons stems from early experiments using the axons of squid preparations (Cole and Curtis, 1939; Young, 1938). The ionic composition of a cell's resting membrane potential and neurons during excitation, which gives rise to the shape and characteristics of transient electrical potentials, were also initially described by studies conducted in the squid axon (Hodgkin and Huxley, 1946, 1952).

The ability to make such discoveries was in part due to the large diameter of the axon. This allowed researchers to place the preparation in defined salines and easily replace the cytoplasm with a known composition of ions while conducting experiments. Wires could be placed inside the large axon to measure and control voltages across the membrane (review in Hodgkin, 1976).

The electrophysiological concepts learned from these early studies were applied to a variety of other cells within animals and plants. Organelles, such as mitochondria, can also be investigated using these methods. The fundamentals of electrical conduction are derived from properties of the various single neurons that make up nerves. Thus, when a nerve is stimulated to produce action potentials from multiple neurons, the resultant additive response within the nerve is referred to as a compound action potential (CAP) (Erlanger et al., 1924).

A CAP is an average of the electrical activity of individual neurons. Groups of neurons have variable conduction velocities, causing the deflections in an extracellular recording to occur at different times. The relative amplitude of the deflections generally correlate with how many neurons are within each grouping (Cragg and Thomas, 1957). The speed of conduction velocity for a particular nerve is primarily based on myelination and axon diameter. The repetitive rates of electrical excitation are a result of the removal of inactivation within voltage-gated sodium channels. It is interesting to note that, even though crustaceans may not use myelination as wrappings around a nerve, there is a form of connective tissue that wraps around their neurons. These wrappings can aid in increasing the rate of conduction for small axon diameters much like myelin in vertebrates (Castelfranco and Hartline, 2015, 2016; Hartline 2008). Many of the ion channels found in plants and animals - such as annelids, insects, crustaceans, and humans - have similar pharmacological profiles. For example, compounds such as Mn^{2+}, Cd^{2+} and 4-aminopyridine (4-AP) and tetraethylammonium (TEA) (Atkins et al., 2021; Pankau et al., 2022; Tanner et al., 2022; Moran et al., 2015; Thiel et al., 2013). Thus, evolution has provided a similar means for different organisms to regulate electrical conduction, excitation, and ionic balance of cells. Participants in a classroom setting can investigate and compare actions of various pharmacological compounds by examining the effects on the CAPs. New pharmacological agents can also be screened for their effects on neuronal function relatively easily and cost-efficient as opposed to using mammals or amphibians.

To gain an understanding of the concepts mentioned above (conduction velocity, refractory period, and the bases of compound action potentials), many teaching modules use nerves from amphibians due to the ease in obtaining the animal and in maintaining the nerve in a minimal saline (i.e. frogs, bullfrogs and even toads). However, there are very set rules for using vertebrates for research and teaching. In the USA, the Institutional Animal Care & Use Committee (IACUC) promotes humane care and use of animal subjects in research. There are federally mandated functions of the IACUC which cover all vertebrate animals. This requires consent by the teaching institution and set procedures to follow along with inspections. Animal use permission for use of lobsters may exist in some European countries and Australia.

In addition, to remove the two sciatic nerves from an amphibian one needs to pith the animal after rendering it unconscious, which can be a difficult task. Generally, only the sciatic nerves from the hind limbs are used from frogs due to the size of the smaller forelimbs. Housing and feeding of these animals, as they only eat moving objects like crickets, requires additional live animals to be managed.

Thus, the goal of this report is to present an alternative animal model, the lobster, from which nerves can be obtained to teach similar educational concepts.

Fewer animals are required, as it is possible to obtain up to 10 or 20 nerves per lobster. The nerves stay viable for many hours in minimal saline. Overall cost and effort in maintaining non-vertebrate animals in a teaching facility are less demanding, depending on the facility and access to animals. There are other invertebrate preparations that exist for teaching labs (crayfish, earthworm, insects, mollusks) that do not offer as many nerves for easy dissection. The crab leg nerves offer some, but not as many, abdominal nerves or a ventral nerve cord as long as the lobster.

In addition to teaching the principles of electrical conduction, the lobster preparation can be used to conduct demonstrations or experiments on physiological concepts prior to dissecting the various nerves out of the animal. Thus, a classroom teaching module in neurobiology can address many concepts with a single lobster. In this protocol, we present recordings from muscles which can be made both extracellularly and intracellularly. The properties of synaptic transmission are readily shown through the use of the abdominal superficial flexor muscles. This preparation demonstrates spontaneous, evoked neural motor nerve activity within an intact ventral nerve cord (VNC). The VNC-induced motor nerve activity is modulated by compounds such as serotonin or octopamine. The compounds are placed on the VNC while recording synaptic responses in the superficial flexor muscles (Baierlein et al., 2011; Kennedy and Takeda, 1965a,b; Strawn et al., 2000).

The nerve to the superficial flexor muscles can be transected to record miniature or spontaneous quantal events. Thus, the quantal hypothesis of synaptic transmission and the mechanistic effects of the neuromodulators can be addressed (Glusman and Kravitz, 1982; Harris-Warrick and Kravitz, 1984; Kupfermann, 1979; Saelinger et al., 2019). Saline of different compositions can be used, such as lowered Na^+ or raised Ca^{2+}, to address effects on synaptic transmission without damaging the future nerve preparations to be used in the same animal. If intracellular recordings cannot be performed in the superficial flexor muscles, the nerve can be transected close to the superficial flexor muscles and recorded with an extracellular suction electrode to address the effects of neuromodulation of the spontaneous evoked motor nerve activity derived within the VNC (Strawn et al., 2000). The nerve activity to the superficial flexor muscles is also driven by sensory stimulation of the cuticle in the same segment in which one is recording; thus, concepts in neural circuitry can be addressed.

Additionally, the concept of sensory proprioception can be demonstrated and experimented with prior to dissecting the nerves. A segmental nerve of the abdomen can be transected close to the VNC and pulled into a suction electrode to record the sensory input from the muscle receptor organ (MRO) in the dorsal aspect of the abdomen (Alexandrowicz, 1951). Thus, the abdominal segment of interest can be flexed and elongated while recording from the nerve. This would demonstrate the correlation of neural activity to the rate of movement of the segments as well as the static position of the joint in segments within the abdomen (Eckert, 1961; Kuffler, 1954; Pasztor and MacMillan, 1990). The nerve being recorded from can also later be used for future experimentation. Isolated nerves from crustaceans are known to live well in minimal saline for several hours. When stored and cooled (4 to 10°C), it is common to have nerves viable the next day for teaching purposes.

PART I
METHODS

1.1. Animal

Live lobsters *Homarus americanus* (Milne-Edwards) measuring 25–30 cm in body length were obtained from a grocery store (local store in Lexington, KY, USA) and housed in a chilled seawater aquarium (Instant Ocean) at 34 ppt and 14 °C-15°C with constant aeration.

1.2. Dissection

The dissection process for obtaining the nerves of the walking limbs and the segmental nerves of the abdominal segments are shown in video format: to accompany this report :

(https://www.youtube.com/watch?v=RvVMsekpgfU).

The segmental nerves are a mix of sensory and motor nerves. The procedures are also shown in Figure 1. Euthanasia of the lobster is performed using large plant pruning shears or scissors with a cut across the cephalothorax, posterior to the eye sockets. It may be easier to start with the eye socket and cut across the head. Then a second cut between the eye stalks to the rostrum. This is to transect the nerves from the subesophageal ganglion to

1

the cerebral ganglion (also termed the supraesophageal ganglion) (Shuranova et al., 2006). The cuticle around the gill chambers is removed along with the gills on both sides of the cephalothorax. The internal organs (i.e., hepatopancreas, stomach and any loose tissue) within the anterior aspect of the cephalothorax chamber can be removed with tweezers and flushed with lobster saline. The flushed saline can be discarded. The saline used for lobster consists of (in mM) 470 NaCl, 7.9 KCl, 15.0 $CaCl_2$-$2H_2O$, 6.98 $MgCl_2$-$6H_2O$, 11.0 dextrose, 5 HEPES acid and 5 HEPES base adjusted to pH 7.5.

Figure 1. **Schematic representation of euthanasia of the lobster for loss of higher function. The green dotted lines show the two main cuts to be made with one across the cephalothorax, posterior to the eye sockets. Although it may be easier to start with the eye socket and cut across the head. A second cut between the eye stalks to the rostrum to split the cerebral ganglion (also termed the supraesophageal ganglion).**

After the animal is euthanized, cut the 1st, 2nd and 3rd walking legs off the animal as close to the base of the leg as possible and place in chilled saline until time to dissect out the nerves. The leg nerves could be dissected and stored in saline to be ready for recording compound action potentials (CAPs). See section 2.4 below for details on and ventral nerve cord (VNC).

PART II
DISSECTIONS AND RECORDINGS
IN THE VARIOUS PREPARATIONS

2.1 Recording Synaptic Activity on the Superficial Flexor Muscle

The spontaneous motor nerve activity to the superficial flexor muscles is maintained best if the ventral nerve cord (VNC) can remain intact as much as possible within the cephalothorax. Thus, after removing the rostral part of the cephalothorax one can remove the intestine and hepatopancreas out the body cavity in the cephalothorax region. One can also remove the cuticle covering the gill chambers and cut off the gills. If the cephalothorax is to also be placed in a saline dish one should rinse the cavity with saline before pinning the preparation in the dish. The abdomen and cephalothorax are pinned in a long dissecting dish with a Sylgard coated bottom. However, leaving the cephalothorax attached does present an issue as a much longer dish and more saline is required than if one just uses the abdomen. The cephalothorax and abdomen can be separated by transecting the VNC above the 1st abdominal rib (A1) and the cephalothorax caudal to the last pair of walking legs. The spontaneous motor nerve activity to the superficial flexor muscles tends to depress in isolated abdominal preparation more rapidly than with the VNC maintained in the cephalothorax (T1-T5 section). The swimmerets need to be cut off on both sides of the abdomen as close to the body as possible to prevent them from moving while making a recording.

The abdomen should be pinned down with the ventral side facing the observer in a Sylgard dish. One can now expose the VNC and the nerve roots as well as the superficial flexor muscle. With a scalpel, cut the articulative membrane (anterior to posterior) along the midline, and then with a scalpel or scissors a cut is made along the rib laterally from the midline to the lateral side (Figures 2 and 3).

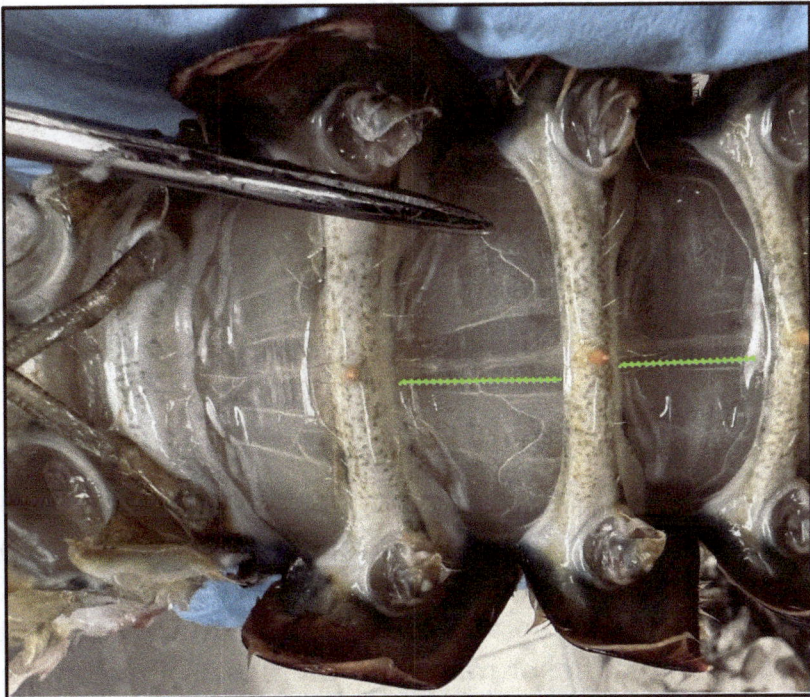

Figure 2. **Exposing the ventral nerve cord and superficial flexor muscle. Using a scalpel, cuts are made from rib to rib through the articulative**

membrane (anterior to posterior) along the midline. Shown as the green dotted lines. The cut needs to be very shallow so as not to cut the ventral nerve cord or the nerve roots of the ventral nerve cord.

To expose the superficial flexor muscle the articulative membrane is cut on the rostral side of the rib (Figure 3A). This flap of articulating membrane is then removed and reflected rostrally so to expose the superficial flexor muscles, VNC and segmental ganglia to the bath solution. (Figure 3B).

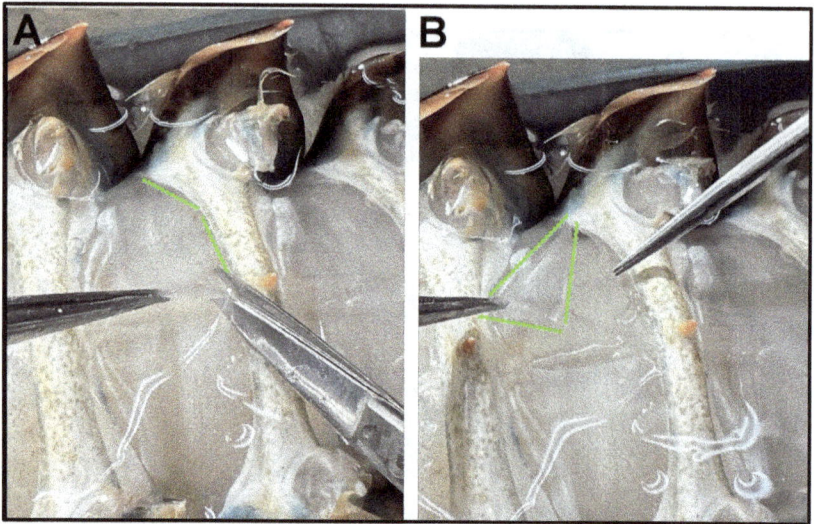

Figure 3. **Exposure of the superficial flexor muscle. (A) The articulative membrane is cut on the rostral side of the rib with care so as not to cut the attachments of the superficial flexor muscle to the ribs as shown with the green dotted line. The articulative membrane can be gently lifted to slip one blade of the scissors under the membrane and above the superficial flexor muscle. (B) This triangular flap (outlined with the**

green dotted lines) of articulating membrane is then removed and reflected rostrally to expose the superficial flexor muscles as pointed to by the tip of the tweezers.

The abdomen is placed in a dish and submerged in lobster saline to cover the muscle. The lobster muscle fibers appear to be slightly more difficult to impale than crayfish muscle due to the connective tissue around muscle bundles. This muscle in crayfish is commonly used as a teaching preparation (Baierlein et al., 2011). To record the spontaneous nerve evoked synaptic events in the muscle fibers, the muscle fiber is impaled with a sharp intracellular electrode (3 M KCl, 20–30 MegaOhm resistance; catalogue # 30-31-0 from FHC, Brunswick, ME, 04011, USA). Excitatory and inhibitory junction potentials (EJPs and IJPs, respectively) can be obtained. As one samples different muscle fibers one will notice different populations in the amplitude of the EJPs and frequency (Figure 4).

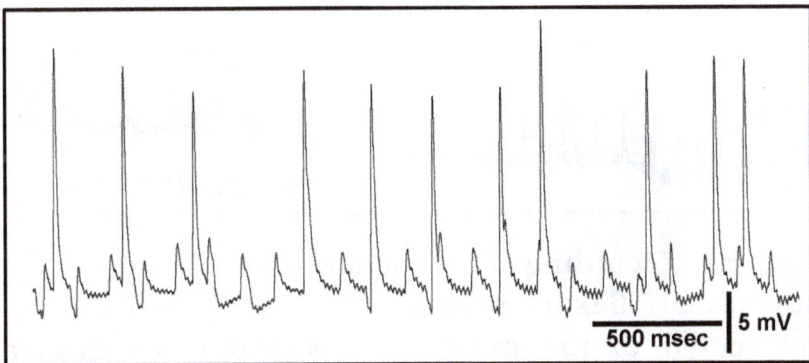

500 msec   5 mV

Figure 4. Synaptic excitatory junction potentials (EJPs) recorded in the superficial flexor muscle. With the 3rd nerve root left intact to the superficial flexor muscle, EJPs can be measured with the intracellular electrode impaled into the muscle fibers. The 3rd root is stimulated by some intrinsic drive with the ventral nerve cord to produce action potentials in the nerve resulting in evoked EJPs in the muscle fibers. The resting membrane potential of the muscle was -62 mV.

To see how a neuromodulator, such as serotonin (5-HT), alters the nerve activity a drop of saline containing 200 microM to 0.5 mM 5-HT made in lobster saline can be placed over the VNC of the segment being recorded (Saelinger et al., 2019; Strawn et al., 2000). The increased muscle contraction may dislodge the intracellular electrode.

The non-evoked, spontaneous quantal events occurring from random fusing of synaptic vesicles of the motor neurons can also be observed by transecting the 3rd root nerve bundle close to the superficial flexor muscle (Figure 5). The responses are highly likely quantal or multi-quantal events and not evoked as the nerve root has been cut and is not electrically stimulated.

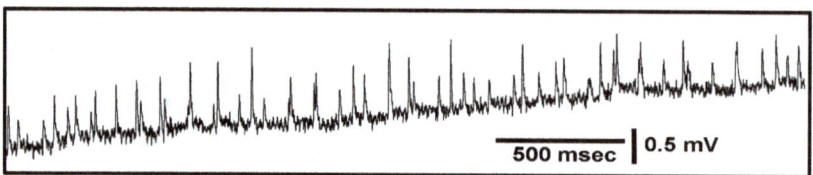

500 msec | 0.5 mV

Figure 5. Recording spontaneous quantal events in the superficial flexor muscle. With the nerve root transected to the muscle the spontaneous evoked events stop occurring and only the spontaneous quantal events will be recorded with an intracellular electrode impaled into a muscle fiber.

If one leaves the longest possible length of the nerve root from the VNC to the superficial flexors then the nerve root can be used to record spikes (extracellular action potentials). The 3rd root has branches from the VNC to the abdominal muscles. The branch of the nerve to the superficial flexor muscles can be pulled into a suction electrode to record nerve activity (Figure 6). To make a suction electrode a step by step detailed video shows the procedures (Baierlein et al., 2011). To ensure a tight fit with the plastic tip of the suction electrode, petroleum jelly (i.e., clear Vaseline) is used to seal the nerve around the suction electrode.

To promote an increase in nerve activity one can, rub the cuticle using a stiff-bristled paintbrush. If the brush is mounted on a micromanipulator there is better control of the movements. A light brushing on the side of the cuticle on the same side one is recording the activity will produce the larger response. The brushing motion will result in the baseline of the signal waving in some preparations. If one is very careful in the brushing motion and the abdomen is well pinned down as not allow the abdomen to move, when recording the EJPs and IJPs with intracellular recordings as mentioned above, the increase in frequency of the EJPs and IJPs can be observed with the brushing motion. Likewise, nerve activity can be altered by exposing the VNC to neuromodulators. Placing a drop of 5-HT on the ganglion (see above) will increase the frequency of spikes with extracellular recordings or of EJPs as measured with intracellular recordings of the muscle fibers.

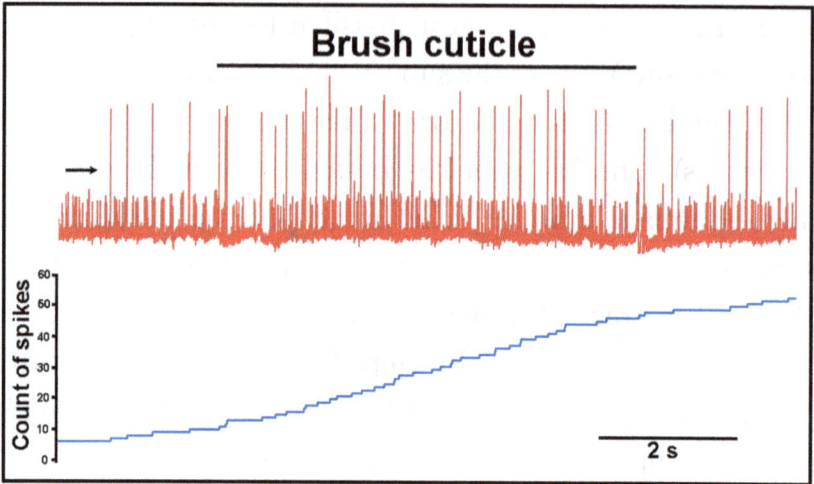

Brush cuticle

Count of spikes

2 s

Figure 6. Spike recording of the 3rd root. The 3rd root recording with a suction electrode provides a means to record the extracellular spikes and effect of stimulating sensory nerves on the cuticle and/or the effect of neuromodulators on the ventral nerve cord. The ipsilateral side of the abdomen was lightly brushed to enhance the frequency of the spikes in the recording. With a window discriminator, set at the level of the arrow, counts of the number of spikes occurring are able to be determined to index the effect of the activating the sensory nerves.

2.2. Activity of the Muscle Receptor Organ (MRO)

One can record the nerve activity from the muscle receptor organ (MRO) in the dissected abdomen as mentioned above. It is relatively easy to do after obtaining recordings from the superficial flexor muscles or the neural activity from the 3rd root.

A segment nerve from the 2nd root off a ganglion from the VNC needs to be cut close to the VNC and the connective tissue around the rib (Figure 7). It will be easiest to use A2 for this recording as the cut end of the rostral part of the abdomen needs to be pinned down to the Sylgard dish or if the cephalothorax is still attached it needs to be pinned down close to the joint between the cephalothorax and abdomen.

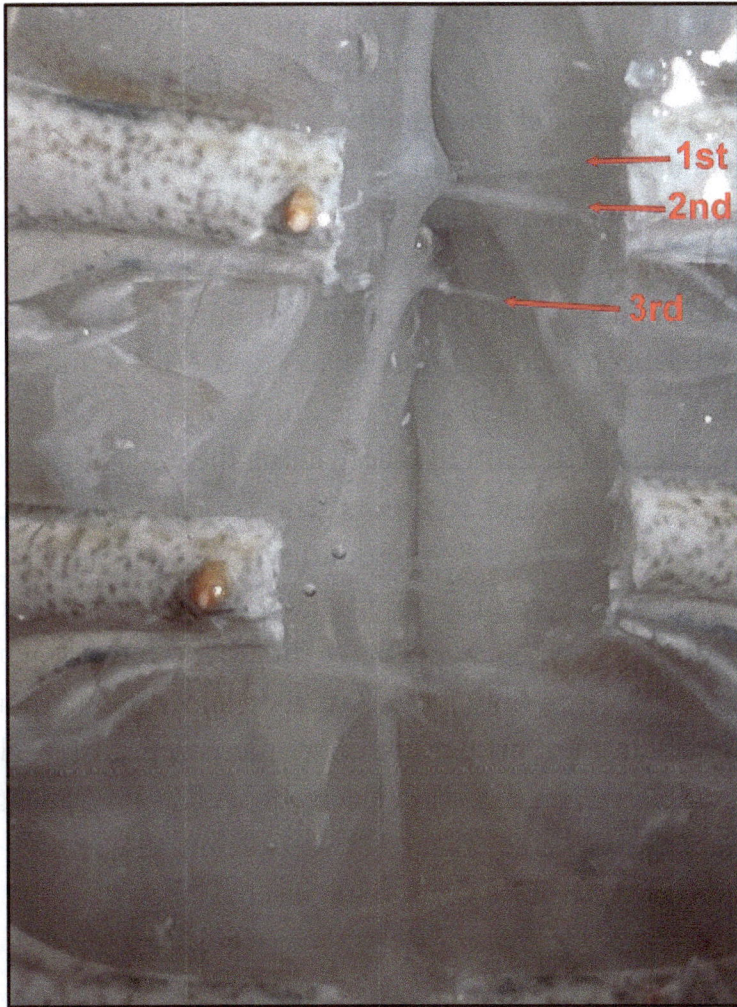

Figure 7. **The nerve roots within an abdominal segment. A section of a rib is removed to expose the**

roots. The 2nd root can be cut close to the VNC to then be pulled into a suction electrode to record the extracellular spikes from the muscle receptor organ (MRO).

To isolate the 2nd root from a ganglion, transect both the 1st and 2nd roots close to the VNC in the segment of choice (i.e., A1 or A2). Carefully cut the cuticle of the rib close to the lateral side of the animal. This is best done by using scissors and cutting through the holes where the swimmerets were attached on both sides at the base of the rib. If the articulating membrane over the superficial flexor muscle was already removed caudal to the rib being removed, then the rib can be turned outward toward the rostral aspect of the abdomen to expose the connective tissue under the rib encasing the 2nd root. The connective tissue needs to be carefully sliced along the underside of the rib to free the 1st and 2nd roots. The 1st and 2nd roots can later be used to record compound action potentials. Thus, care should be used here not to pinch the nerves so further recordings can be obtained. Once the rib is removed, gently slide the scissors under the 2nd root and trim off nerve branches and connective tissue to these ventral muscles. Pull the nerve into a suction electrode with as long a length of nerve as possible (Figure 8). Petroleum jelly is used to seal the nerve around the suction electrode.

Figure 8. Recording of the afferent nerve activity from the muscle receptor organ (MRO). The activity from the sensory neurons of the MRO can be monitored with a suction electrode recording of the 2nd root. Removing the muscle of the flexor aids in pulling a longer length of nerve into the suction electrode so it will not slip out while flexing the telson to stimulate the MRO proprioception.

Recording the nerve activity while holding the telson of the abdomen and slowly bending it (flexing) should cause the MRO to produce neural activity related to static positions (i.e., the degree of flexing the joint being monitored). If the rostral part of the preparation is pinned down well, then the abdomen can be flexed at relatively faster rates to see how activity correlates with the rate of flexing the joint. Two large spikes of different amplitudes are usually observed. The spikes with similar amplitudes relates to static positions of the joint and the other group of spikes with a different amplitude relate to dynamic movements of the joint.

Smaller amplitude spikes may also be observed which are likely from sensory neurons which detect movement/pressure/hair deflections on the outside of the cuticle while moving the abdomen. Figure 9A shows the responses of the MRO nerve from a slow flexion of the abdomen while Figure 9B shows the responses in a more rapid flexion. A recording with a stable trace can be obtained if one isolated the segment with pins. For demonstration of the MRO recording, without damaging the other nerves for isolation, the rest of the abdomen is not pinned down. The baseline in the trace moves with disturbances in the saline bath while flexing the abdomen. There are many references related to the anatomy and physiology of the MRO preparation in both lobsters and crayfish which can readily be obtained by students in a course. A review of the MRO literature and video format on how to perform MRO recordings in a crayfish are presented in Leksrisawat et al., (2010) and recordings in crayfish and lobster (Kuffler, 1954).

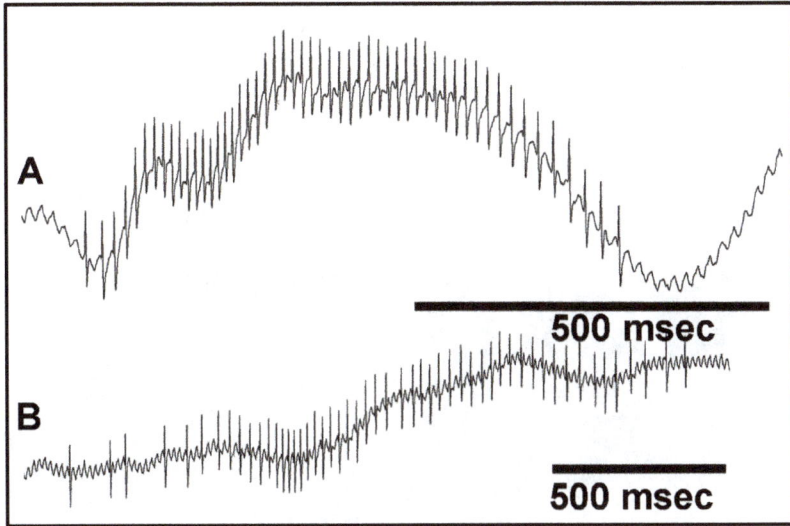

Figure 9. Afferent nerve recording of the MRO during flexion of the abdomen. (A) Slowly flexing the abdomen and (B) rapidly flexing the abdomen will produce differing frequency of spikes. The baseline movement in the recording will occur when moving the large abdomen and disturbing the saline bath.

2.3 Isolating the VNC and Segmental Nerves

We found two different dissection approaches which can be used to isolate the segmental nerves from the abdomen. Both approaches may take a sample lobster to first work out what approach is best for the person doing the dissections. With a beta test lobster, one can use methylene blue to aid in tracing the nerve roots as one dissects around the lateral sides Figure 10. It is easy to lose the nerve as it drops between the muscle fibers. Thus, with staining one can learn how the segmental nerve transverses through the muscles and along the connective tissue of the ribs to the lateral side of the animal (Wiersma and Hughes, 1961). Once the segmental nerve is isolated from the large flexor muscles it is easier to see the nerves proceeding to the smaller dorsal extensor muscles. The nerve can be cut at the end when it branches to the extensor muscles.

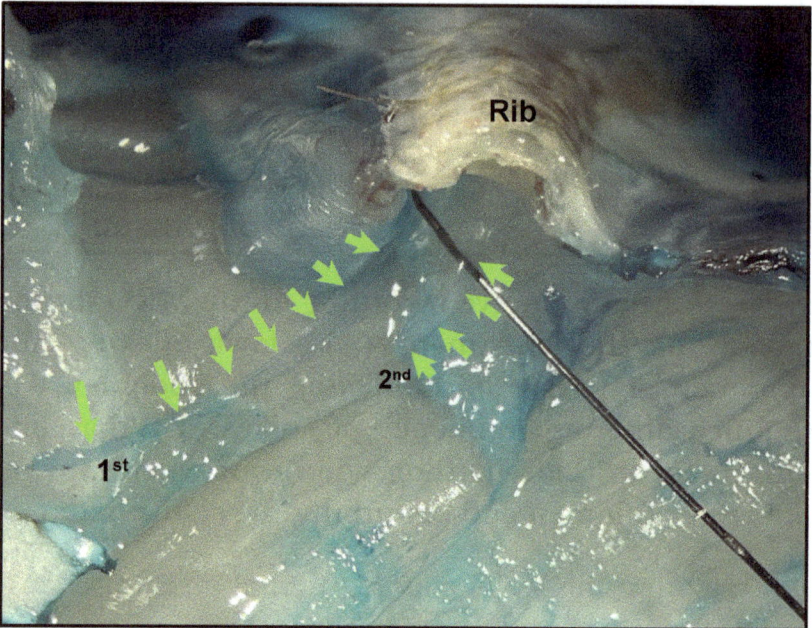

Figure 10. **Isolating the segmental nerves of the abdomen. Methylene blue staining is used to aid in the identification of the anatomy in deciding an approach for dissection of the abdominal nerves on a sample lobster. A cross-section of a rib is facing the viewer. The 1st and 2nd roots off the ventral nerve cord run close to the rib within the segment. Connective tissue by the rib initially encases the nerves and needs to be carefully trimmed away.**

One approach when not using a stain is to tie a thin string around the nerve prior to cutting it away from VNC. The nerve can be transected close to the VNC. The string can be used to gently lift the nerve while dissecting the connective tissues and nerve branches while trying to obtain the longest segmental nerve possible. Unraveling non-waxed dental floss into individual strings provides strings that work well for this procedure. The damaged ends can later be cut away with the string before pulling them into a suction electrode or they can be left on the nerve and used for helping to place the nerve over the wires when using a nerve chamber recording device such as those obtained from ADI (ADI instruments). If one does not use a string, a serrated pair of tweezers may aid in holding on to the end of the nerve while gently lifting to isolate the nerve from the muscles. Likewise, the damaged end can be cleanly trimmed prior to physiological recordings.

In isolating the segmental nerves there is a lot of flexor muscle to be removed and connective tissue to be trimmed away while following the segmental nerves to the lateral sides. The 1st root of each ganglion curves slightly rostral within the segment while the 2nd root proceeds with a slight caudal direction while approaching the lateral side. The 1st roots are typically shorter in length than the 2nd roots. We found starting with the A1 segment and working towards the caudal end and removing the nerves as we progressed to be a good approach. The cavity of the abdomen may need to be rinsed with fresh saline to remove debris when proceeding with the dissection.

As one works along one side of the abdomen, the other side can be followed in the same manner. If the VNC was not pinched or damaged in the dissection, then it can be removed as well after all the segmental nerves are removed. This provides another preparation to record CAPs. Dissecting the segmental nerves on both sides of the abdomen for A1–A5 takes about 1 hour as it is required to be patient while following the nerves and removing pieces of the flexor muscle in each segment.

Another approach that provides more freedom to move the abdomen around and dissect with the lateral side pinned down in a dish, is to transect the abdomen into two lateral halves. We preferred this approach as then two people can isolate the segmental nerves in two different dishes. To use this approach the abdomen is isolated from the cephalothorax. The telson below the last rib and caudal to the 6th abdominal ganglion is removed. A small section of each rib next to each ganglion is removed. The scissors can cut the articulating membrane between each rib, keeping on the midline, while proceeding to each rib. After each rib has a small section removed and the articulating membrane cut, the lateral sides can be gently pulled away from each other exposing the VNC and the roots branching away from the VNC (Figure 11).

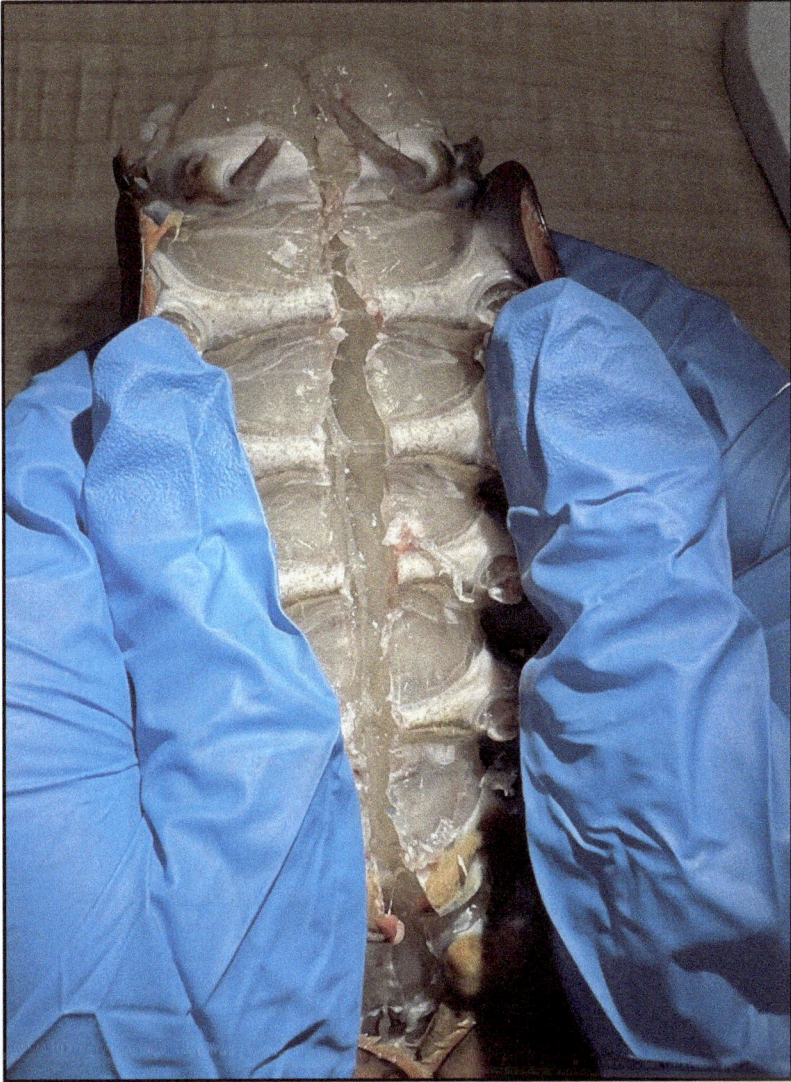

Figure 11. The isolated abdomen is split to isolate the segmental nerves and ventral nerve cord. The segmental nerves (1st and 2nd roots) are able to be dissected for each segment on both sides providing 20 nerve preparations in addition to the ventral nerve cord of the abdomen.

The length of the VNC from A1 past A6 can be isolated by transecting the roots along the length of the VNC (Figure 12). It is easy to miss the 3rd root as it has branches with some proceeding to deeper muscles in the abdomen. Pinching with tweezers the VNC by the rostral end from the A1 ganglion, the VNC can be lifted while ensuring each root is transected.

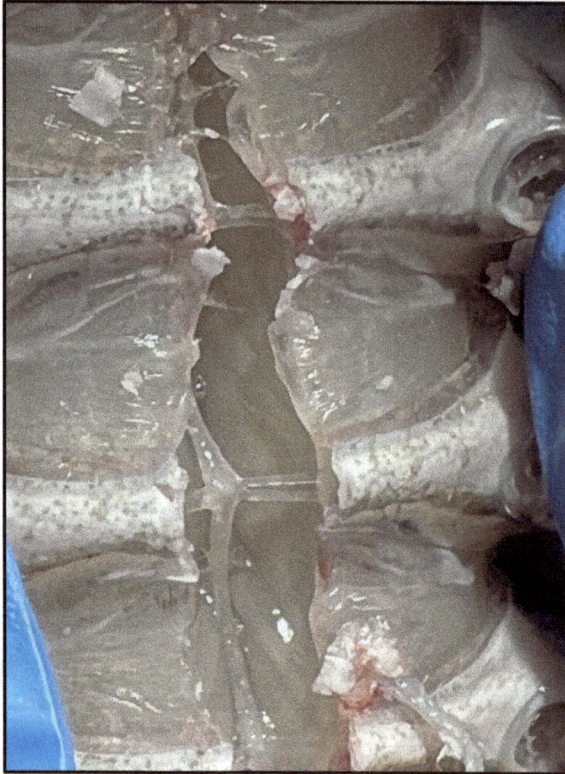

Figure 12. **Exposing the nerve roots from the ventral nerve cord. The root can be cut along the ventral nerve cord and the ventral nerve cord isolated from the abdomen.**

After the VNC is removed and placed in saline, the abdominal muscles can be cut along the midline to the dorsal cuticle. The abdomen can then be turned over and the cuticle cut along the dorsal middle to produce the two halves. The two haves can be pinned to a Sylgard coated dish and the roots carefully dissected away from the connectives tissue around the ribs and muscles as mentioned above (Figure 13).

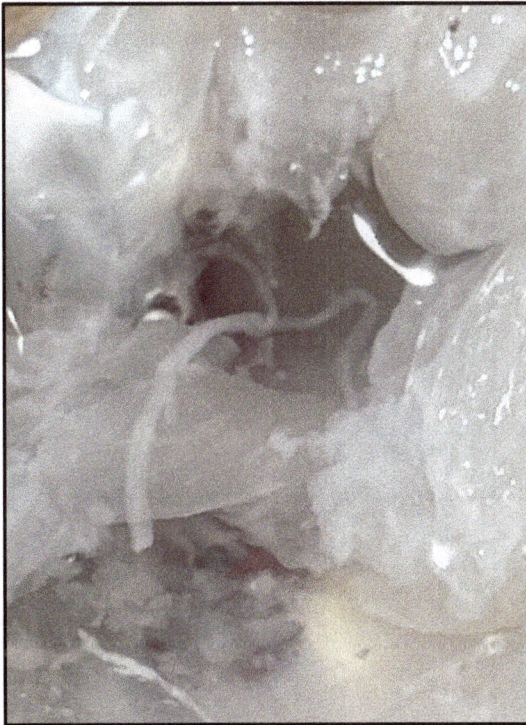

Figure 13. **The segmental nerve in the abdomen is isolated from the rib and the surrounding muscle. The nerve is to be transected at the most distal location along the dorsal side of the abdomen and removed.**

2.4 Recording the Compound Action Potentials of Nerves

Place the leg on a paper towel for ease in dissection. With a scalpel or scissors, cut along the dorsal edge from the base to the distal propus. The next cut is along the ventral length of the leg (Figure 14). This needs to be done with care as the nerve is closer to the ventral side of the leg. At each segment, carefully cut through to the next segment. Once the dorsal and ventral cuts are made the cuticle can be carefully peeled away on one side and with a scalpel cutting any tissue adhering to the cuticle on the side being removed.

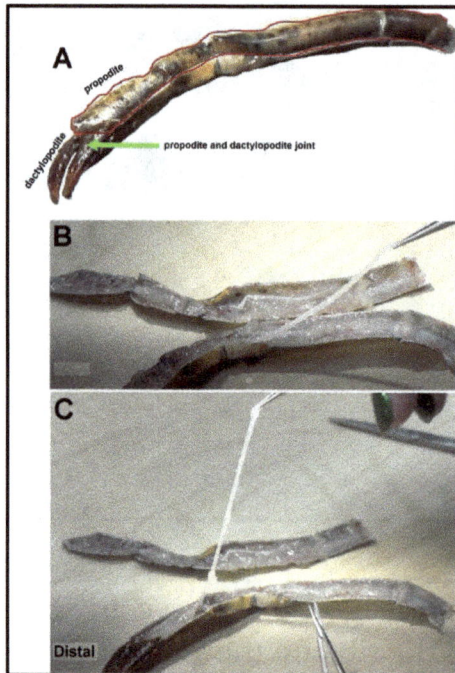

Figure 14. **Isolating the leg nerve. (A) The dissection of the leg to expose the main leg nerve. Cut along the**

ventral and the dorsal sides of the leg with a scalpel as shown in red. (B) Removal of the lateral side of the leg to expose the main leg nerve. The main leg nerve is carefully dissected out of the leg. (C) The nerve can be cut at the distal end of the propodite before entering the dactylopodite (the last distal segment on the leg).

The nerve can now be separated from the muscle on the side facing the observer. The nerve will be attached with connective tissue at the joints and at the base of the leg. It may be best to pick the nerve up with serrated tweezers at the very base of the leg and begin peeling it distally while cutting branches off the nerve which innervate the remaining muscles of the leg and include sensory nerves from the apodeme/cuticle. Once the nerve is peeled to the distal end at the propodite and dactylopodite joint, it can be transected and placed in a dish of saline. The damaged end where the tweezers gripped the nerve can be cleanly cut with sharp fine scissors. This process is repeated for all three pairs of walking legs which provides six long nerves. To obtain the rest of the nerves is mentioned below.

To record the compound action potentials of the leg nerves, the segmental nerves and the isolated VNC, each can be placed in a Petri dish containing physiological lobster saline and used at individual set ups in a classroom. Two different approaches are commonly used. One with a suction electrode on each end of the nerve where one is for stimulating and the other for recording. To ensure a tight fit with the plastic tip of the suction electrode, petroleum jelly (i.e., clear Vaseline) is used to seal the nerve around the suction electrode (Figure 15).

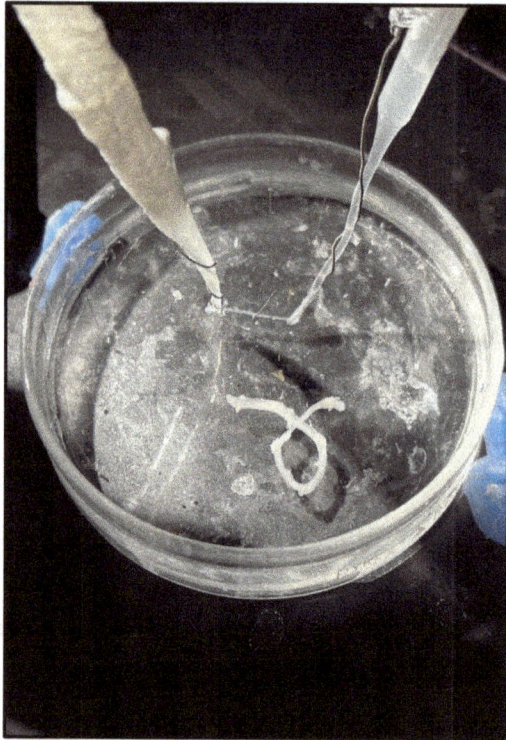

Figure 15. **Recording set up to measure the compound action potential. A segmental nerve from the abdomen of the lobster is ready for recording with both ends placed into suction electrodes and a small amount of petroleum jelly (Vaseline) around the nerve to ensure a seal with the nerve ends of the electrodes. Both suction electrodes contain saline in contact with the silver wire inside prior to pulling the nerve into the electrode. The ventral nerve cord of the lobster is also in the dish for keeping it moist in saline. The dish is coated with Sylgard so insect pins can be used to stabilize the nerves in the dish.**

Another approach is to use the commercially available nerve chamber, commonly used for frog/toad sciatic nerve recordings (Figure 16). Teaching protocols provided by AD Instruments (ADInstruments, Inc.; 4360 Arrowswest Drive, Colorado Springs, CO. 80907, USA) have standard software programs which can be used to stimulate and record CAPs as well as measure refractory periods in the nerve. For the slow conducting neurons, the standard sciatic nerve software parameters may need to be modified to longer time periods to catch the full waveform.

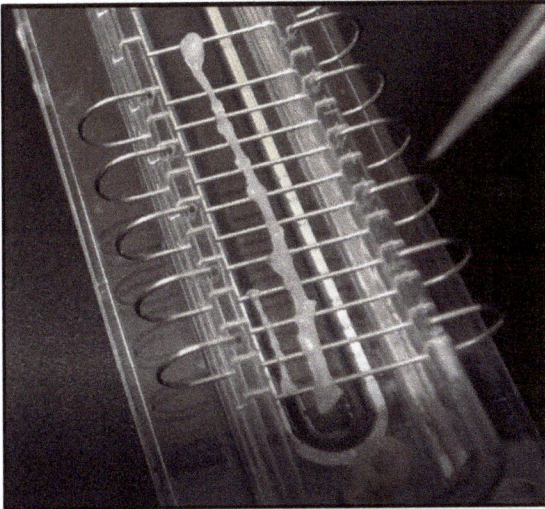

Figure 16. **A standard commercially available nerve chamber for stimulating and recording compound action potentials of sciatic nerves such as those obtained from a frog/toad. This chamber is one sold by AD Instruments. A segmental nerve of the abdominal segment (smaller one) and the ventral nerve cord are shown.**

Common physiological measures in nerve recordings are to teach recruitment of neurons to a maximum CAP amplitude and to obtain conduction velocities (Figure 17). Measuring the length of the nerve from the recording electrode to the stimulating electrode obtains the length of the nerve. Measuring the time from the stimulus artifact in the electrical recording to the peak of the various deflections in the CAP on can determine the conduction velocity for the subsets of neurons. The absolute and relative refractory periods with twin pulse stimulation are depicted in Figure 18. Manipulations in experimental procedures can also be applied such as the effect of saline temperature, pharmacological agents (i.e., TEA, 4-AP) and alterations in ionic compounds in the saline (i.e., raised K^+, lowered Ca^{2+}) (Atkins et al., 2021; Tanner et al., 2022).

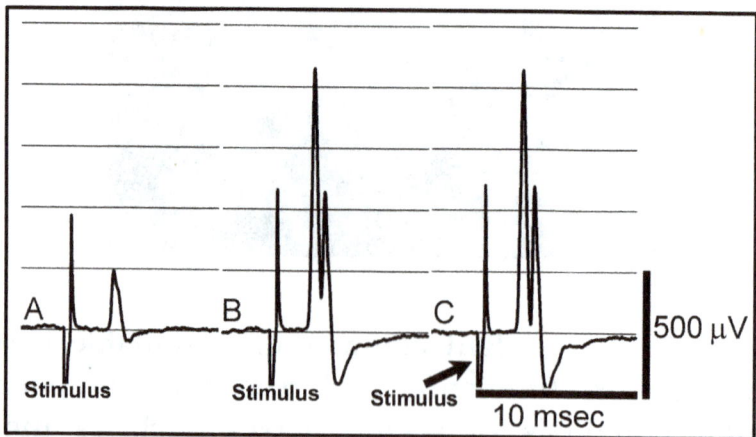

Figure 17. **Recruitment of neurons to produce a compound action potential (CAP) from the main nerve in a leg or a segmental abdominal nerve of the lobster. The stimulation voltage is increased until a**

maximum amplitude and shape of the **CAP** is obtained. The rapid initial deflection in the trace is from the stimulus which travels in the saline bath to the recording electrode. The deflection following the stimulus response is the **CAP** which is larger from **A** to **C** as the stimulus was increased.

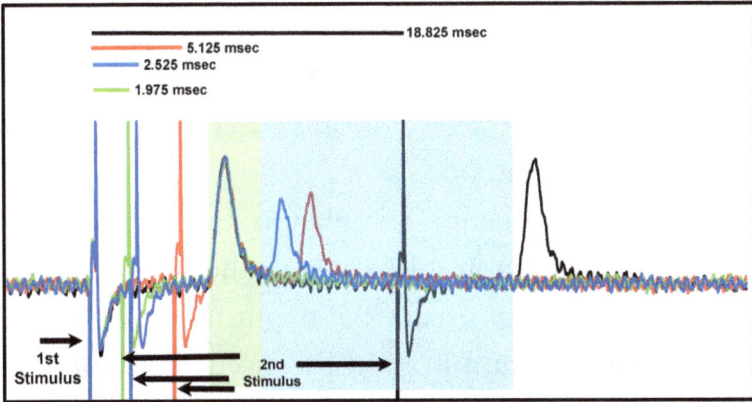

Figure 18. **Refractory periods of neurons. Providing two stimulus pulses with varying delay allows one to determine absolute and relative refractory periods of the nerves. As shown with two stimuli with varying delay, the absolute refractory period of this nerve was 2.525 msec and full recovery the compound action potential (CAP) occurred at 18.825 msec (relative refractory from 2.525 to 18.825 msec). The bars above represent the timing between the 1st stimulus and the 2nd stimulus. The 1st CAPs are superimposed, and 2nd CAPs are shown at varying times away from the 1st . Note the 2nd CAP during the absolute refractory period is not present.**

With trains of constant stimulation with increasing voltage one can denote the repetitive recruitment of the

CAPs as more neurons are recruited. In keeping the voltage constant and varying the time from the 1st stimulation to the 2nd stimulation one can determine absolute and relative refractory periods. The time from the 1st stimulation to the 2nd when no CAP is observed as the 2nd stimulus is given with a longer delay from the 1st is considered an absolute refractory period. The time to when a small CAP is observed to a fully recovered amplitude of the CAP which was induced by the 2nd stimulation is referred to as the relative refractory period.

Upon high frequency stimulation the production of CAPs becomes compromised where the CAP might not occur at all, or a very small CAP might be present (Figure 19). This phenomenon is good for discussions for a class and relates to what might happen in the CNS of mammals during an epileptic seizure.

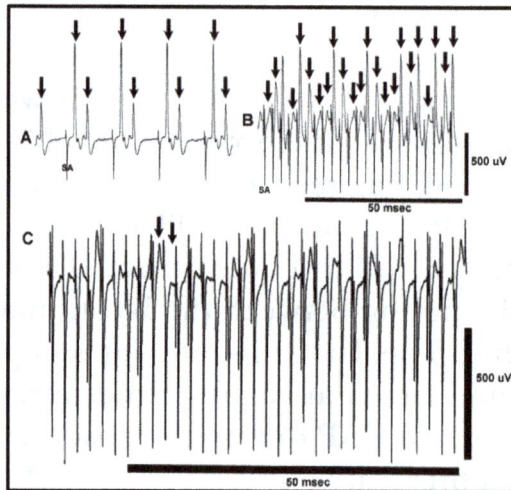

Figure 19. **Repetitive stimulation produces varied responses due to complex compound action potential**

(CAP) shapes and various refractory periods. With continuous stimulation at various frequencies, the amplitudes of the CAPs vary due to some neurons having different refractory periods. (A) There is a complex CAP with two different groups of neurons producing the CAP. (B) As the frequency of stimulation increases the double peaked CAP is not observed with each stimulation likely because of some neurons are in absolute refractory. (C) With even a higher rate of stimulation, the neurons within the nerve are either in absolute refractory or relative refractory as the amplitude of the CAPs are small and sometimes a CAP does not occur (i.e. as shown with the 2nd arrow in C). The stimulus artifact (SA) occurs due the electrical event from the electrical signal from the stimulating electrode traveling in the saline bath to the recording electrode. The SA is followed by the electrical events induced by the nerve. The arrows point to where the CAPs are present or where they should be occurring.

Varying the external concentration of the ions in the saline can produce unexpected results which can produce interesting discussions for a class. Also, conducting primary literature searches on such topics can aid in scientific literacy. One topic which has resulted in lively discussions is the effects on neuronal activity with hypocalcemia. The phenomenon of nerves showing heightened activity with lowered calcium concentration is illustrated in Figure 20. The bath is switched out to a saline with no Ca^{2+} added and the nerve is repetitively stimulated at a low frequency or 0.5 Hz (1 time every 2 seconds) with a maximum amplitude of the CAP. What is generally observed is that the nerve will randomly start producing CAPs. Sometimes varying amplitudes of the CAPs occur, and the frequency of the occurrences increases and then can stop altogether except for the CAP being evoked by the background low frequency stimulation from the experimenter induced stimulation.

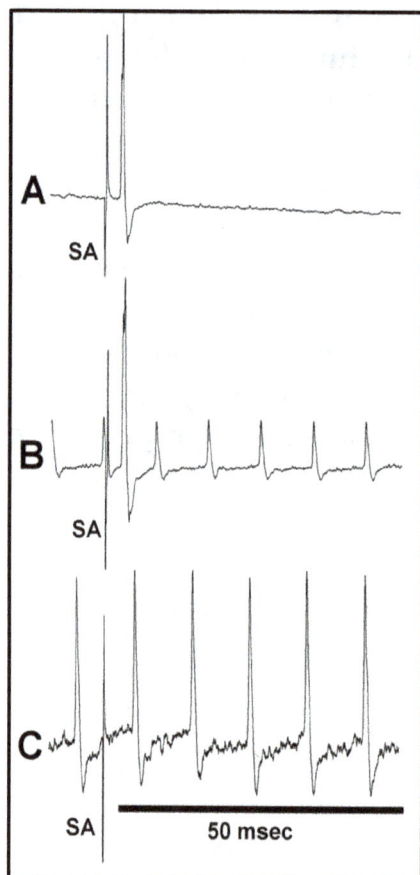

Figure 20. **Low extracellular Ca²⁺ effects on spontaneous activity of neurons. The stimulus artifact (SA) from the stimulating electrode induces the**

evoked CAP while the other CAPs just start appearing randomly due to the low extracellular Ca^{2+} concentration. (A) Initial evoked CAP by stimulating the nerve with a suction electrode. The stimulus artifact (SA) is followed by the electrical event induced by the nerve producing a CAP. (B) After 15 minutes in a bathing solution without Ca^{2+} the nerve will start to produce random CAPs. (C) After 20 minutes a higher frequency of the CAPs start to occur in addition to the experimenter induced evoked CAP.

Generally, the isolated nerves of freshwater crayfish and marine crabs and lobsters are viable after 24 hours in saline when stored in a common refrigerator 4-6°C immersed in the appropriate saline. However, the function may be somewhat compromised and not as robust as they were initially after dissection. To illustrate this for the lobster preparation the nerve of a leg was isolated and stimulated to produce a CAP right after dissection and to repeat the stimulation 5 and 23 hours later (Figure 21). The dish with the preparation was stored in a refrigeration and set up again for stimulating and recording at various times. Thus, the stimulus artifact (SA) may appear different due to the way the nerves are placed in the suction electrode and sealed with petroleum jelly. The nerve in the preparation used to obtain the responses shown in Figure 21 produced an extra CAP from the stimulation after being maintained for 23 hours.

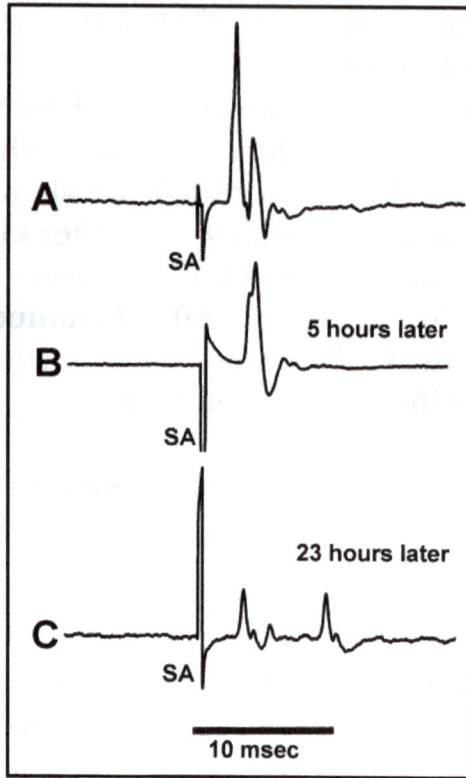

Figure 21. The isolated nerves of the lobster remain viable for hours in saline. The compound action potential (CAP) can be elicited in the nerve after a day in saline. The nerve was repetitively sampled to induce CAPs after being isolated from the leg, 5 hours and 23 hours. Each time the nerve was arranged with the suction electrodes and removed from the electrodes to be maintained in saline within a refrigerator (4–5 °C).

PART III
DISCUSSION

The goal of this report is to provide a protocol and organism for teaching concepts in animal physiology and neurophysiological laboratories. As compared to the use of sciatic nerves in amphibians, overall expense and the number of animals needed are reduced. Some of the same principles of neurophysiology taught using the frog sciatic nerves can be taught using lobster nerves, as they are relatively easy to dissect, and many are available from one animal. The walking legs of commercially sold lobsters are generally long and the main leg nerve can be readily dissected out, normally providing 6 long nerves from the first 3 pairs of walking legs in about 30 minutes of dissection time. The ventral nerve cord of the lobster is also able to be removed rapidly and is close to the length of the abdomen. The segmental nerves of the abdomen take longer and are more tedious to dissect but with a little practice one becomes proficient and can obtain at least 10 or more nerves. In addition, the concepts of synaptic transmission at the NMJs, sensory perceptions as well as synaptic integration of sensory-VNC- motor nerve and modulation of the neural circuitry can be covered all within a signal animal. Measuring the CAPs along the VNC can also provide an intriguing class discussion based on the anatomy of the VNC (Furshpan and Potter, 1959; Watanabe and Grundfest, 1961). Students can compare

CAPs obtained with the nerves of the leg and lateral segmental nerves to the VNC.

Other commonly obtained marine crustacean preparations in the North America, such as the relatively large crabs (*Cancer magister, Callinectes sapidus*), can be used to obtain relatively long nerves from the walking legs (Brock et al., 2023; Elliott et al., 2023). The Blue crab (*Callinectes sapidus*), commonly obtained along the Gulf of Mexico , around Florida and up the east coast of the US to Maine, as well as the Dungeness crab (*Cancer magister*), which is native to the northern coast of California to Canada, are good preparations for obtaining nerves from the walking legs. The nerve dissection from the crab legs are detailed in earlier reports (Brock et al., 2023; Elliott et al., 2023).

One caveat with using a lobster is the ability of the animal to tail flip even after being euthanized. The sharp points on the underside of the abdomen can cut a hand even when wearing surgical gloves. In addition, one needs to keep the rubber band in place over the claws until the chelipeds are removed and not to place one's hand or fingers in from the mouth parts as they can also produce a nasty pinch. Of course, if one does not dissect them right after buying live lobsters or marine crabs one has to store them in a chilled seawater aquarium. Lobsters and particularly the Dungeness crab survive better in chilled seawater (10 °C). If one is not experimenting with the chelipeds of the animals then they can be kept cold and cooked later for consumption as well as the remaining cephalothorax.

Hopefully, the procedures and results presented herein provide examples of what can be obtained and used for teaching purposes. Prior to using the nerves, other experiments with the intact animal are able to be performed such as behavioral experiments in habituation in the tail flip (Edwards et al., 1999; Pagé and Cooper, 2004), monitoring heart rate and even the rate of respiration by monitoring the beating frequency of the scaphognathite (Mendelson, 1971; Bierbower and Cooper, 2009; Shuranova et al., 2003) with simply recording by two wires just placed under the cuticle by the organs of interest (Listerman et al., 2000). Even electromyograms (EMGs) of the skeletal muscles in the walking legs or chelipeds are readily able to be recorded to demonstrate the correlation between movements and electrical activity (Bradacs et al., 1997).

REFERENCES

Alexandrowicz, J.S. Muscle receptor organs in the abdomen of Homarus vulgaris and Palinurus vulgaris. *Q. J. Microsc. Sci.* **1952**, *92*, 163-199.

Atkins, D.E.; Bosh, K.L.; Breakfield, G.W.; Daniels, S.E.; Devore, M.J.; et al. The effect of calcium ions on mechanosensation and neuronal activity in proprioceptive neurons. *NeuroSci* **2021**, *2*, 353-371.

Baierlein, B.; Thurow, A.L.; Atwood, H.L.; Cooper, R.L. Membrane potentials, synaptic responses, neuronal circuitry, neuromodulation and muscle histology using the crayfish: student laboratory exercises. *J. Vis. Exp.* **2011**, *47*, 2322.

Bierbower, S.M.; Cooper, R.L. Measures of heart and ventilatory rates in freely moving crayfish. *J. Vis. Exp.* **2009**, *32*, 1594.

Bradacs, H.; Cooper, R.L.; Msghina, M.; Atwood, H.L. Differential physiology and morphology of phasic and tonic motor axons in a crayfish limb extensor muscle. *J. Exp. Biol.* **1997**, *200*, 677-691.

Brock, K.E.; Elliott, E.R.; Taul, A.C.; Asadipooya, A.; Bocook, D.; Burnette, T.; et al. The effects of lithium on proprioceptive sensory function and nerve conduction. *NeuroSci* **2023**, *4*(4), 280-295.

Castelfranco, A.M.; Hartline, D.K. The evolution of vertebrate and invertebrate myelin: a theoretical computational study. *J. Comput. Neurosci.* **2015**, *38*(3), 521-538.

Castelfranco, A.M.; Hartline, D.K. Evolution of rapid nerve conduction. *Brain Res.* **2016**, *1641*(Pt A), 11-33.

Cole, K.S.; Curtis, H.J. Electric impedance of the squid giant axon during activity. *J. Gen. Physiol.* **1939**, *22*, 649–670.

Cragg, B.G.; Thomas, P.K. The relationship between conduction velocity and the diameter and internodal length of peripheral nerve fibers. *J. Physiol.* **1957**, *136*, 606-614.

Eckert, R.O. Reflex relationships of the abdominal stretch receptors of the crayfish. I. Feedback inhibition of the receptors. *J. Cell. Comp.Physiol.* **1961**, *57*, 149-162.

Edwards, D.H.; Heitler, W.J.; Krasne, F.B. Fifty years of a command neuron: the neurobiology of escape behavior in the crayfish. *Trends Neurosci.* **1999**, *22*(4), 153-161.

Elliott, E.R.; Brock, K.E.; Taul, A.C.; Asadipooya, A.; Bocook, D.; Burnette, T.; et al. The effects of zinc on proprioceptive sensory function and nerve conduction. *NeuroSci* **2023**, *4*(4), 305-318.

Erlanger, J.; Gasser, H.S.; Bishop, G.H. The compound nature of the action current of nerves as disclosed by the cathode ray oscillograph. *Amer. J. Physiol.* **1924**, *70*, 624-666.

Furshpan, E.J.; Potter, D.D. Transmission at the giant motor synapses of the crayfish. *J. Physiol.* **1959**, *145*(2), 289-325.

Glusman, S.; Kravitz, E.A. The action of serotonin on the excitatory nerve terminals in lobster nerve-muscle preparations. *J. Physiol.* **1982**, *325*, 223–241.

Harris-Warrick, R.M.; Kravitz, E.A. Cellular mechanisms for modulation of posture by octopamine and serotonin in the lobster. *J. Neurosci.* **1984**, *4*, 1976–1993.

Hartline, D.K. What is myelin? *Neuron Glia Biol.* **2008**, *4*(2), 153-163.

Hodgkin, A.L. Chance and design in electrophysiology: an informal account of certain experiments on nerve carried out between 1934 and 1952. *J. Physiol.* **1976**, *263*(1), 1-21.

Hodgkin, A.; Huxley, A. Potassium leakage from an active nerve fibre. *Nature* **1946**, *158*, 376–377 (1946).

Hodgkin, A.L.; Huxley, A.F. A quantitative description of membrane current and its application to conduction and excitation in nerve. *J. Physiol.* **1952**, *117*(4), 500-544.

Kennedy, D.; Takeda, K. Reflex control of abdominal flexor muscles in the crayfish: the twitchsystem. *J. Exp. Biol.* **1965a**, *43*, 211–227.

Kennedy, D.; Takeda, K. Reflex control of the abdominal flexor in the crayfish: the tonic system. *J. Exp. Biol.* **1965b**, *43*, 229–246.

Kuffler, S.W. Mechanisms of activation and motor control of stretch receptors in lobster and crayfish. *J. Neurophysiol.* **1954**, *17*, 558- 574.

Kupfermann, I. Moduatory actions of neurotransmitters. *Ann. Rev. Neurosci.* **1979**, *2*, 447–465.

Leksrisawat, B.; Cooper, A.S.; Gilberts, A.B.; Cooper, R.L. Muscle receptor organs in the crayfish abdomen: a student laboratory exercise in proprioception. *J. Vis. Exp.* **2010**, *45*, e2323.

Listerman, L.; Deskins, J.; Bradacs, H.; Cooper, R.L. Measures of heart rate during social interactions in crayfish and effects of 5-HT. *Comp. Biochem. Physiol. A.* **2000**, *125*, 251-264.

Mendelson, M. Oscillator neurons in crustacean ganglia. *Science* **1971**,*171*, 1170-1173.

Moran, Y.; Barzilai, M.G.; Liebeskind, B.J.; Zakon, H.H. Evolution of voltage-gated ion channels at the emergence of Metazoa. *J. Exp. Biol.* **2015**, *218*(Pt 4), 515-525.

Pagé, M.P.; Cooper, R.L. Novelty stress and reproductive state alters responsiveness to sensory stimuli and 5-HT neuromodulation in crayfish. *Comp. Biochem. Physiol. A.* **2004**, *139*(2), 149-158.

Pankau, C.; Nadolski, J.; Tanner, H.; Cryer, C.; Di Girolamo, J.; Haddad, C.; et al. Effects of manganese on physiological processes in Drosophila, crab and crayfish: Cardiac, neural and behavioral assays. *Comp. Biochem. Physiol. C* **2022**, *251*, 109209.

Pasztor, V.M.; MacMillan, D.L. The actions of proctolin, octopamine and serotonin on the crustacean proprioceptors show species and neurone specificity. *J. Exp. Biol.* **1990**, *152*, 485-504.

Saelinger, C.M.; McNabb, M.C.; McNair, R.; Bierbower, S.; Cooper, R.L. Effects of bacterial endotoxin (LPS) on the cardiac function, neuromuscular transmission and sensory-CNS-motor nerve circuit: A crustacean model. *Comp. Biochem. Physiol. A.* **2019**, *237*, 110557.

Shuranova, Z.P.; Burmistrov, Y.M.; Cooper, R.L. Bioelectric field potentials of the ventilatory muscles in the crayfish. *Comp. Biochem. Physiol. A.* **2003**, *134*, 461-469.

Shuranova, Z.P.; Burmistrov, Y.M.; Strawn, J.R.; Cooper, R.L. Evidence for an autonomic nervous system in decapod crustaceans. *Interl. J. Zool. Res.* **2006**, *2*(3), 242-283.

Strawn, J.R.; Neckameyer, W.S.; Cooper, R.L. The effects of 5-HT on sensory neurons, central, and motor neurons driving the abdominal superficial flexor muscles in the crayfish. *Comp. Biochem. Physiol. B* **2000**, *127*, 533-550. (See Erratum 128:377-378, 2001 for missing 1/2 of figure).

Tanner, H.; Atkins, D.E.; Bosh, K.L.; Breakfield, G.W.; Daniels, S.E.; Devore, M.J.; et al. The effect of TEA and 4-AP on primary sensory neurons in a crustacean model. *J. Pharmacol. Toxicol.* **2022**, *17*, 14-27.

Thiel, G.; Moroni, A.; Blanc, G.; Van Etten, J.L. Potassium ion channels: could they have evolved from viruses? *Plant Physiol.* **2013**, *162*(3), 1215-1224.

Watanabe, A.; Grundfest, H. Impulse propagation at the septal and commissural junctions of crayfish lateral giant axons. J. Gen. Physiol. **1961**, *45*, 267-308.

Wiersma, C.A.G.; Hughes, G.M. On the functional anatomy of neuronal units in the abdominal cord of the crayfish, Procambarus clarkii. *J. Comp. Neurol.* **1961**, *116*, 209-228.

Young, J.Z. The functioning of the giant nerve fibres of the squid. *J. Exp. Biol.* **1938**, *15* (2), 170–185.

FUNDING

This research was funded by Beckman Scholarship (K.E.B.) and personal funds (R.L.C.).

ACKNOWLEDGMENTS

We thank Mrs. Shea Carr, Assistant Professor at Centre College, Danville, Kentucky, for editorial assistance and scientific evaluation of this protocol.

We thank Karen Mireau Rimmer for editing and helping to process this text for publication.

ABOUT THE AUTHORS

Kaitlyn E. Brock

Kaitlyn E. Brock is an undergraduate student at the University of Kentucky pursuing a dual degree. She will graduate in May 2025 with a B.S. in Neuroscience and a B.S. in Psychology. She first became interested in research when she sat in on one of Dr. Robin Cooper's classes and immediately wanted to become involved. She joined Dr. Cooper's lab shortly afterwards and has been working with him for three years on various neurophysiology projects and teaching protocols.

Currently a Beckman Scholar, Kaitlyn serves as an Undergraduate Research Ambassador. She has had the opportunity to present her research at many conferences across the country and intends to pursue a Ph.D. in neuroscience following graduation and focus her career on research.

A native of Lexington, Kentucky, Kaitlyn is an avid reader and hiker. She enjoys playing soccer, listening to true crime podcasts, and spending time with friends and family.

ABOUT THE AUTHORS

Robin L. Cooper

Dr. Robin L. Cooper obtained a dual B.S. in Chemistry and Zoology from Texas Tech University in 1983 and obtained a Ph.D. in 1989 in Physiology from the School of Medicine.

He then went on for postdoctoral training (1989-1992) at the University of Basel, School of Medicine, Basel, Switzerland and a second postdoctoral stint (1992-1996) in the Department of Physiology at the University of Toronto, School of Medicine, Toronto, Canada.

In 1996, Dr. Cooper joined the Department of Biology at the University of Kentucky and is now a Professor. He also obtained a BSN in nursing in 2012 and practiced as an RN from 2011 to 2017. He has received several teaching awards over the years and continues to mentor students in research based activities and assist them in publishing for peer review.

In his spare time, he Zooms and Facetimes with his first grandchild, Rose, and cycles the backroads of Kentucky.

To Contact the Authors
please email:

Kaitlyn E. Brock
kaitlynbrock@uky.edu
Robin L. Cooper
RLCOOP1@uky.edu

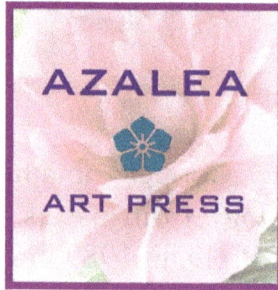

AZALEA

ART PRESS

To contact the publisher
please email:
KarenMireauBooks
@gmail.com

For print book and Ebook orders
please visit:
www.Lulu.com

www.ingramcontent.com/pod-product-compliance
Lightning Source LLC
Chambersburg PA
CBHW071343290326
41933CB00040B/2118